The History of Quantum Physics

All rights reserved

First edition 2017
Copyright © 2017 M. Melih Gördesli
www.melih-goerdesli.at

Cover design using
photographs by Erlend Davidson
photo "Quantum Hydrogen on Graphene"
CC BY SA Lizenz - via www.flickr.com

Published using Amazon CreateSpace
ISBN 978-1-5430 -3152-8

Edited by Shaydon B. Ramey

M. MELIH GÖRDESLI

The History of Quantum Physics

There was a time when the newspapers said that only twelve men understood the theory of relativity. I do not believe there ever was such a time. On the other hand, I think I can safely say that nobody understands quantum mechanics.

Richard P. Feynman

Dedications

In the following dedication, I would like to take a moment to list and describe those people who have inspired me with their love of science, which has always stood up to ethnic or capitalist repression.

Ernst Abbe (1840 – 1905)

Ernst Abbe, mathematician and physicist, grew up in a very simple working-class family.

His extraordinary achievements in school led him to receive the financial support of his father's employer. He in return was to work as an accountant in the company. However, this did not happen. Ernst Abbe contin-

ued his achievements and became a university professor. He contributed to the clear improvement of the microscope by taking into account the wave properties of light in his calculations. In 1876, Abbe suffered from typhoid fever and was unable to work for several weeks, causing him financial difficulties. His employer Carl Zeiss took the opportunity to make him a partner in his company. There, he quickly developed further improvements to the microscope. Ernst Abbe became a millionaire, but one with a social conscience. He was of the opinion that the capital generated by the individual must flow back into the general public. After the death of Carl Zeiss, he became the sole head and changed the company into a cooperative. Abbe introduced the eight-hour working day and sick pay. He also shared the company's profits with every employee.

Nikola Tesla (1856 – 1943)

Nikola Tesla, inventor, engineer and futurist, was born into a Serbian family, the father of which was a strict orthodox priest. He studied mathematics and physics in Graz and later philosophy in Prague with his uncle's financial support. His career as an electrical engineer began at a telephone company in Budapest. Before he left for America, he worked on novel dynamo motors in Paris. However, his first prototype, an induction motor, did not arouse interest in Europe, whereupon he finally left America to work for Thomas Edison. After a disagreement that came to be part of the so-called War of Currents, Tesla left Edison and worked for a year as a road construction worker due to his poor financial situation. He took his ideas to two businessmen and founded a company with them, but he was deceived and later declared bankruptcy. He did not give up and registered his first patents in 1886. In the same year, Tesla signed a contract with Westinghouse Electric Co. and won the War of Currents against Edison. Compared to Edison's direct current, Tesla's alternating current proved to be a much more efficient technology. Countless technologies still in use today (electric motors, generators, radio controls, etc.) hark back to Nikola Tesla.

Martin Karplus (born in 1930)

Martin Karplus, a theoretical chemist and Nobel Prize Laureate (2013), was born as the child of a Jewish family in Vienna. Karplus was to become a doctor like his two famous grandfathers, since the Jews in this area faced the least discrimination. When his family fled the National Socialists to America, these plans were put on hold. This dramatic escape from their Austrian homeland likely played a large role in shaping the young Karplus. In the summary of his fifty-page biography, he even mentions that this traumatizing experience plays an essential role in his worldview and in his approach to science. One day, when his brother was given a chemistry kit and started experimenting, Karplus became curious and insisted on conducting experiments. His parents forbade him to do so - they thought it was a bad combination for two teenagers to be running around with explosive materials. Instead, Karplus got a microscope, which was decisive for his interest in the natural sciences, particularly biology.

If quantum mechanics hasn't profoundly shocked you,
you haven't understood it yet.

Niels Bohr

Foreword

The scientist's task is essentially to research nature and explain its processes. At first, this may sound banal and self-evident, but not every phenomenon we encounter on a daily basis and take for granted can be grasped by the man on the street upon closer examination. In particular, once we move from the macrocosm to the microcosm, the interpretation of these phenomena becomes even more comprehensive and more difficult. The processes of nature may seem trivial to us, because our understanding thereof has become a familiar part of our existence, and essentially always has been. Why the sky appears blue, why the earth is round and the sun rises and sets, why water is liquid and evaporates at high temperatures, or where the vibrant colors of a rainbow come from – we rarely question or even notice all of this.

These issues, which at first glance appear to be simplistic, concerned ancient philosophers long ago. Whether Zeno of Elea, with his divisional paradox, wherein he described the relationship of space, time, and motion, or Democritus with his "indivisible" atom, which later proved to be decomposable, or Plato's parables for the theory of knowledge – all of these and the philosophical ideas of other free thinkers form the basis for the description of natural processes. To what extent these considerations, with their metaphysical character, influenced the history of physics, particularly quantum physics, and contributed (and continue to contribute) to their further development is just one of the many questions that this work will attempt to answer.

There is no doubt that philosophers, natural scientists[1], and artists have made fundamental contributions to the positive development of human civilization. Although not all of them now enjoy social recognition, ironically enough, these free spirits have made technical and scientific progress possible. It must be emphasized, however, as history also shows, that the most important contributions came from academics, aristocrats, and other elitist environments, which simply had the right connections and suffered no lack of time or money. All the more admirable are the contributions to social progress made by figures from more modest social conditions. Natural scientists and philosophers are not only those who have a degree in their field of study, but also those who are passionate about it and who see their possibilities for development in it.

In his book "Einsteins Schleier," Austrian quantum physicist Anton Zeilinger writes about amateur physicist Thomas Young, whose simple double-slit experiment turned the world upside down:

---

[1] Mathematicians and computer scientists included

"Perhaps it is the case that someone who is outside the subject, has his own independent income, and is therefore not dependent on whether or not he is recognized as a professional colleague can more easily take unusual steps in a completely new direction than one whose career as a physicist directly depends on the opinions of his peers."

The present work draws attention to the most important principles of quantum theory – far removed from mathematical confusion and complex ideas. For, as is well known, the art of efficient enlightenment lies in breaking things down into their essential parts and in the simplicity of language.

Table of Contents

INTRODUCTION	17
THE NATURE OF LIGHT	21
Historical Development	21
Particle or Wave?	24
THE BIRTH OF THE FIRST QUANTA	31
The Ultraviolet Catastrophe	32
Einstein's Photo Effect	36
WAVE-PARTICLE DUALISM	41
BOHR'S ATOMIC MODEL	47
Atoms – The Invisible World	49
THE MATTER WAVE	53
De Broglie's Bold Idea	54
Schrödinger's Wave Equation	57

HEISENBERG'S UNCERTAINTY PRINCIPLE   62
ZENO'S ARROW PARADOX   64

THE COPENHAGEN INTERPRETATION   69

THE FIFTH SOLVAY CONFERENCE ON "ELECTRONS AND
PHOTONS"   72
THE EPR PARADOX   76
DE BROGLIE-BOHM THEORY   80
A QUANTUM MOUSE IN SCHRÖDINGER'S CRATE   89

EPILOGUE   91

BIBLIOGRAPHY   93

Introduction

Quantum theory has existed for more than a century. Like any other theory, it has also had to submit to strict examination. Since its successful introduction, it has mainly been referred to as "quantum physics" or "quantum mechanics." Today, alongside the theory of relativity, it represents a cornerstone of modern physics.

When Max Planck launched the first "quantum," he could not have known the serious consequences his extraordinary discovery would have for classical physics. Newton's laws lost their importance for the atomic order of magnitude at once; it was realized that the world of atoms does not work according to our mechanical worldview. The planetary model of the atom, where electrons circle around the nucleus, proved to be false. Even matter did not seem to be what it should be. Light, electrons, protons, neutrons, etc. demonstrated both the properties of a wave as well as those of a particle. There was then talk of particle-wave dualism and matter was called a matter-wave.

Even accuracy no longer mattered. Werner Heisenberg realized that it is impossible to determine two measurements of a material wave - such as speed and location - at the same time. Instead, his deep insight into reality shows that its foundation is based on probabilities, so-called wave functions, to use Erwin Schrödinger's words, which collapse each time we observe them according to the Copenhagen interpretation.

To this day, it has not been completely clarified what is going on in the world invisible to the naked eye. It is no secret that our insights about atomic scale are based entirely on mathematical principles. They are mostly calculations that allow us to penetrate unknown areas without destroying the atomic structure with external influences such as observation.

Although the nature of the macrocosm up to the microcosm can be adequately described by mathematical and experimental means for everyday purposes, there is no guarantee that our environment actually is what we think it is. The work of mathematician Kurt Gödel most clearly shows what thin ice science really is walking on. With his incompleteness theories, he proves that the attempt to build a complete and consistent mathematical system is doomed forever. He also added that there will always be questions that mathematics can not answer. It is therefore uncertain to what extent reality corresponds to the truth. Richard Feynman noted that the paradox is actually the conflict that takes place between reality and our perception of what should be reality.

Quantum physics is certainly not an easy task. Even famous physicists and Nobel Laureates like Albert Einstein, Richard Feynman, and many others admit that the world is obviously not at all like our minds allow it to be. Albert Einstein was right when he said, "The world cannot be so crazy. Today we know the world is so crazy."

If you think you understand quantum mechanics,
you don't understand quantum mechanics.

Richard P. Feynman

The Nature of Light

Before we dive into quantum theory, we first must deal with the nature of light, which we will take as our starting point. We divide, so to speak, the complex nature of quantum theory into small portions and arrange them so that an intelligible plausibility is built up piece by piece.

Historical Development

In Antiquity, perhaps even much earlier, philosophers were already concerned with the phenomena of light. It was assumed at first that light, after hitting the eye, leaves it again as "visual rays" and feels the environment like hands. In addition, the Greek mathematician Euclid of Alexandria posited the linearity of light beams, while his colleagues Heron and Damianos continued to believe that they travel on the shortest path from the eye to the object.

Today, we know that light reflected off an object comes directly into the eye and not the other way around. It is thus only possible to capture the visible through the reflection of light - unless the object itself is a source of light[2].

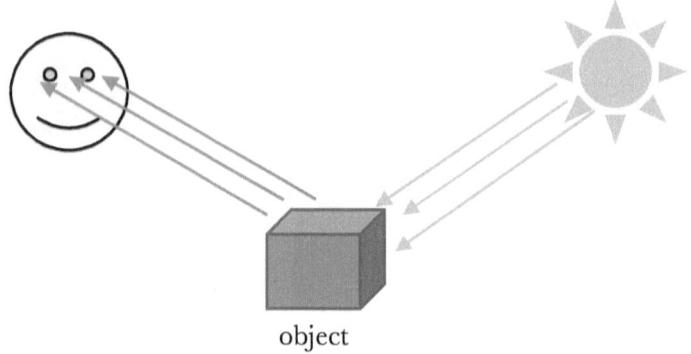
object

It was inevitable that discussions about the speed of light would also arise, and two schools of thought were quickly formed: the one advocated finitude and the other infinitude. For human perception, light is infinitely fast, since far-away light sources such as stars seem to be captured instantly by the eye. But the proponents of the finite speed of light did not stop defending their point of view - light had to move through space somehow and travel a long distance from the sun. The Greek philosopher Empedocles also believed in this finite speed of light. Persian philosophers shared this view as well, as did scholars like Avicenna[3] and Alhazen[4].

---

[2] e.g. candle, lamp
[3] Full name: Abū Alī al-Husain ibn Abdullāh ibn Sīnā
[4] Full name: Abū Alī al-Hasan ibn al-Haitham

Unsurprisingly, these blind convictions failed to form a unanimous opinion on the speed of light. This split would continue for a very long time, until the first concrete evidence shed light on the matter.

The phenomenon of light, which raised many more questions, piqued everyone's curiosity. It spread like a wildfire among philosophers and natural scientists. In time the followers of the finite speed of light were overrun by the Aristotelian theory that the speed of light is infinite. Italian scientist Galileo Galilei was one of the first to attempt a measurement with a coarse experimental setup - but to no avail.

Today we know from the observations and experiments of natural scientists Ole Rømer, Hippolyte Fizeau, and Leon Foucault[5] that light does indeed travel at a finite speed, about 300,000 kilometers per second. However, this is exclusively the case in a vacuum. In the air, it is about 0.3 percent less, in water about 25 percent, and in a glass, the speed of light is slowed down to 47 percent depending on the refractive index. Light reaches its full development, so to speak, only in an airless space. As early as the beginning of the seventeenth century, German scientist Johannes Kepler assumed the dependence of the speed of light on the medium traversed - a theory that proved to be true.

We can even prove the postulates of Euclid, Heron, and Damianos that "visual rays" spread straight along the shortest path with today's cutting-edge technology – however, not as it was then assumed from the eye to the object, but rather the other way around. In his famous lecture series on quan-

---

[5] Leon Foucault was known for proving Earth's rotation using a pendulum.

tum electrodynamics, Richard Feynman explains that light looks for the path that takes the least time.

It is astonishing that freethinking minds from Antiquity without any resources came to such realizations "only" through philosophical thinking. The daily observable natural phenomenon light was still full of surprises and gave further incentive to find out more about its nature.

Particle or Wave?

The nature of light was not entirely clear. It still hid secrets that scientists and philosophers tried to uncover. One of the biggest questions concerned its composition. Since it was known that light spreads at a finite speed, it was initially assumed to have the character of a particle. At the beginning of the 17th century, English scientist and founder of classical mechanics Sir Isaac Newton attempted to explain its propagation with his corpuscular theory, according to which light consists of small particles (so-called "corpuscles"). He let light beams hit a triangular glass prism, revealing the typical rainbow spectrum. He was able to explain the vivid colors by the fact that light consists of a stream of small particles, each with its own color. However, it was assumed that these colors somehow belonged to the prism, whereupon Newton turned the rainbow beam onto a second prism, and behold: white, natural light came out of the prism again.

Newton's corpuscular theory also explains the refraction and reflection of light. Nevertheless, many naturalists disagreed with him, such as Dutch astronomer and physicist

Christiaan Huygens, who clearly stated in his treatise on light that the phenomenon of refraction and reflection can also be described by waves. More precise proof was to be provided in 1912 by German physicist Max von Laue and his colleagues Walter Friedrich and Paul Knipping by means of X-rays projected on crystals. Before this, however, a simple experiment by an ophthalmologist would finally undo Newton's worldview.

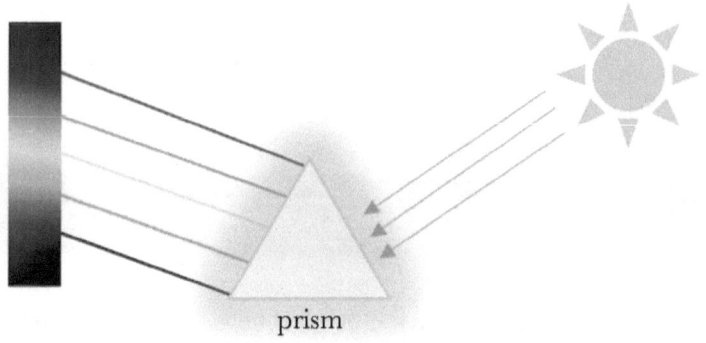
prism

When doctor and hobby physicist Thomas Young happened to be watching two swimming ducks on a pond, he saw the waves they left behind crashing into each other and forming a new pattern. In doing so, he noticed that when two wave crests crashed into one another, they produced a larger crest, and two wave troughs formed an even deeper trough. On the other hand, if a wave trough hit a wave crest and the other way around, the two waves were canceled out.

Two wave crests meet to form a larger wave:

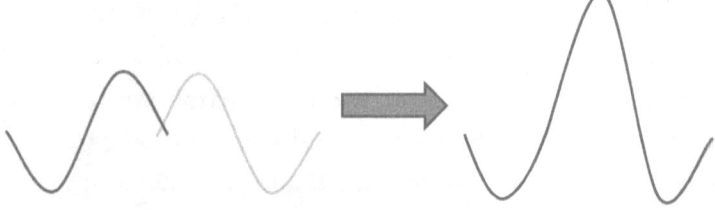

Two wave troughs meet to form a deeper trough:

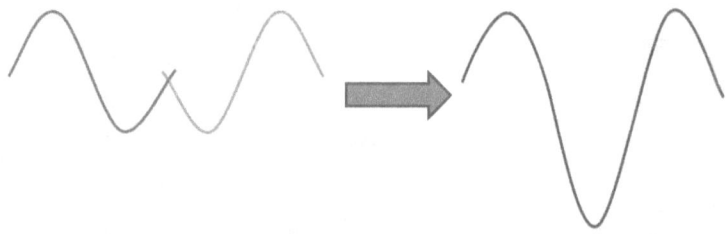

Wave trough meets wave crest and vice-versa - both are canceled out:

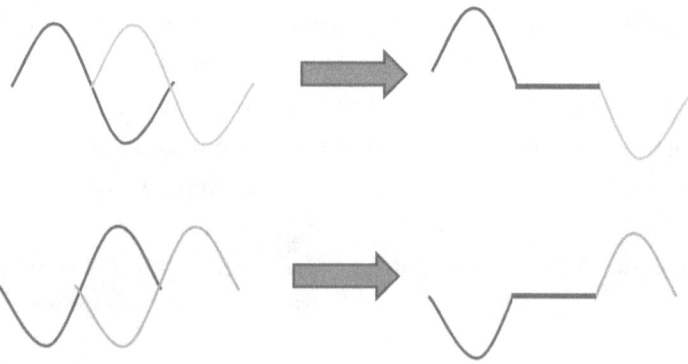

Thomas Young suddenly remembered his old experiment, which was later to go down in history as the famous "double-slit experiment."

Young had installed a partition with two thin gaps between a light source and a screen. Following Newton's corpuscular theory, he expected two bright lines on the screen, as the light particles could only have passed through the two slits. Surprisingly, he discovered an interference pattern[6] instead. Now that he had watched the rippling waves of the ducks, he knew how difficult the striped pattern[7] was to explain.

At first he assumed that light spread like a water wave. Passing through the two gaps, it split into two smaller waves, which interfered with each other. At the point where two waves crests or two wave troughs overlapped, a bright strip appeared on the screen. The dark spots, on the other hand, arose from the collision of wave crests and wave troughs, which canceled each other. Light seemed to behave like a wave.

---

[6] A pattern of light and dark stripes next to each other
[7] Known as diffraction

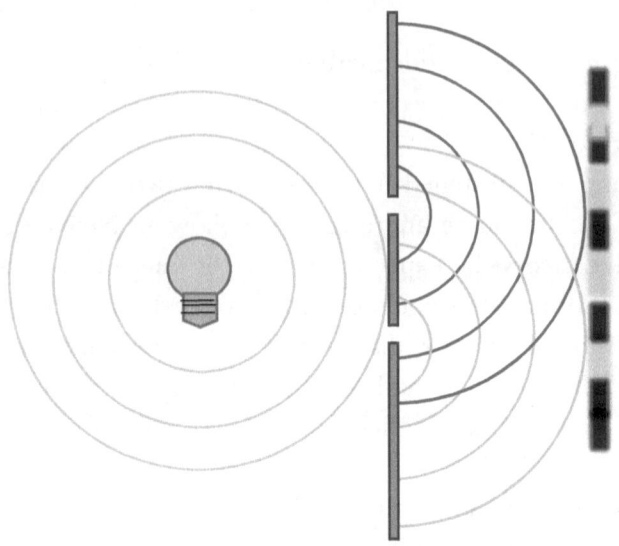

The simple double-slit experiment

The bands from the interference pattern have different light intensities. In the middle, where more waves of the same type meet, a higher intensity can be observed, while the dark spots are caused by the cancellation of the wave crests and troughs.

In 1846, the well-known English experimental physicist Michael Faraday demonstrated the connection between light and magnetism. Almost twenty years later, the Scottish physicist James Clerk Maxwell took the next step and summarized magnetism with electricity. This resulted in electromagnetism, which spreads in the form of a wave. Maxwell was able to prove this mathematically. The interesting thing was that the propagation speed corresponded exactly to the speed of light. Maxwell therefore assumed that light consists of electromagnetic waves.

Newton's corpuscular theory was able to hold its own for a century. With the electromagnetic light theory, almost every light phenomenon seemed to be explained. However, a new, subsequent discovery was to pull the rug out from under scientists' feet and revitalize the particle-wave discussion.

Thomas Young

The Birth of the First Quanta

In the 17th century, Sir Isaac Newton united Galileo Galilei and Johannes Kepler's research on acceleration and planetary motion in his magnum opus *Principia Mathematica* - the law of gravitation was born. The three well-known fundamental laws of motion also hark back to this work. Newton not only laid the foundations for mechanics, but also created the classic Newtonian world image, in which many things could finally be explained logically. Apples falling from trees or planets orbiting around the sun were no longer a mystery but facts that could be adequately described with Newton's laws. From then on, the world was ticking like complex clockwork that was deterministic and causal. Everything was descriptive and calculable. However, two hundred years later, at the beginning of the twentieth century, a German physicist's discovery would completely rock the Newtonian world.

Sir Isaac Newton

The Ultraviolet Catastrophe

Max Planck, the father of quantum physics, probably did not suspect the fundamental significance his decision to study physics would have for the all of natural science. The German physicist had been completely indecisive, and swayed between music, ancient philology, and physics. In spite of the suggestion of a close professor to study something else, since everything fundamental had already been discovered in physics, Max Planck could not be swayed. He successfully completed his studies and then devoted himself to research.

Planck first dealt with the so-called "black body." Physicists had observed that by adding heat to objects, different elements possess different color spectra. They had their own fingerprint, so to speak, just like natural light.

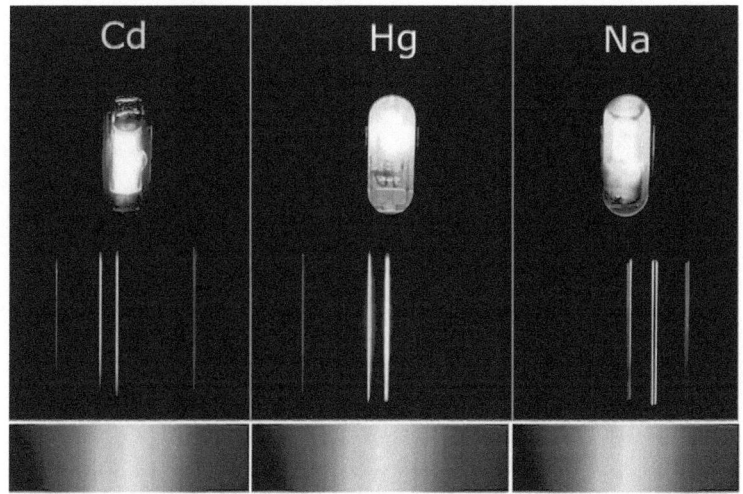

Low-pressure metal vapor lamps with associated emission spectra for cadmium, mercury, sodium

In order to investigate this phenomenon without outside disturbances, a black body was constructed that allowed the almost complete absorption of light. Planck used a cavity that led the light through a tiny hole into its interior and reflected it on the walls. It was found that the color and intensity of the respective material depend exclusively on the temperature. At absolute zero, the objects do not emit radiation. If heat energy is added so that the material begins to glow, it is released again as light energy through the tiny opening.

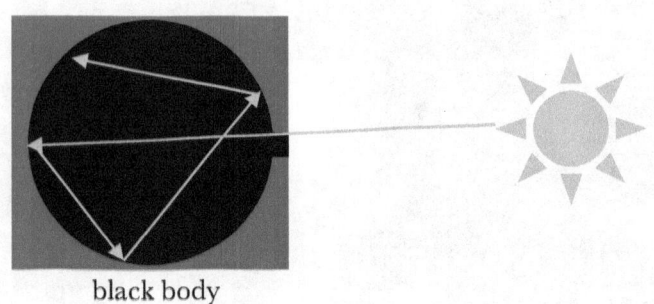

black body

Scientists were unsuccessfully trying to describe this phenomenon with mathematical formulas. The problem with the whole issue - which went down in history as an "ultraviolet catastrophe" - was that according to calculations, heated objects would have to be invisible as soon as their temperatures reached the ultraviolet range. But when a piece of metal is gradually heated, it does not disappear, of course, but its color shifts from glowing red, orange, then yellow, and eventually ends up being white. This could not be explained with classical physics. The previous mathematical equations of Wilhelm Wien and John William Strutt were able to describe only a part of the radiation spectrum.

When Planck looked at the existing equations, he realized that certain particles of radiation energy were always left behind. He initially kept a cool head and tried to tackle this problem with theoretical approaches. According to Planck, there should be some physical explanation for this. But he ended up in an "act of despair," as he described it, until he went back to the work of German physicist Heinrich Hertz.

Hertz had developed an oscillator[8] for the measurement of electromagnetic waves and their emission and absorption. It was earlier assumed that electromagnetic radiation exists in the black body in the form of standing waves. A standing wave can be imagined as the string of a guitar fixed at both ends and vibrating up and down. Planck assumed that these waves could not accept any energy value. Instead, he very carefully formulated that "the energy could be quantized," i.e. that their transmission in packets takes place not in just any size, but rather in a specific one. A blue standing wave, for example, has 3 eV (electronvolt), 6 eV, or 9 eV of energy - but never values in between. Because the oscillation of the waves depends on their frequency (v), he related them as a logical conclusion to the energy (E) of the radiation. Another size, a natural constant, which had escaped all other researchers, completed Planck's mathematical description. Planck could scarcely believe it when he compared the results of his mathematical formula with the phenomenon of the "ultraviolet catastrophe": the problem was miraculously solved!

$$E = h \cdot v$$

The inconceivably small natural constant "h" was given the designation "Planck's quantum of action." Planck had calculated $\mathbf{6.885 \cdot 10^{-34} J \cdot s}$ as a value. That was only four

---

[8] An oscillator is an electronic circuit that produces a periodic, oscillating electronic signal

percent less than the value of **$6.62 \cdot 10^{-34} J \cdot s$** that we know today thanks to technical progress.

At the beginning of the 20th century, Max Planck finally presented his findings to the public. At that time, he still did not realize what a great movement he had set in motion, which was approaching the foundations of classical physics in a purposeful and unstoppable way.

Max Planck

Einstein's Photo Effect

Despite the publication of Planck's results on the heat radiation of black bodies, no one had recognized their fundamental importance. The formula with which Planck had finally remedied the seemingly impregnable ultraviolet catastrophe was seen as a mathematical trick until Albert Einstein, physicist and later Nobel Prize Laureate (1922), was able to explain a known but unexplained phenomenon using Planck's work.

Even before Einstein, natural scientists had observed that electrons were emitted from a metal plate when they were irradiated with light. If the light source was moved closer to the plate, more electrons were released from the plate, but the exit velocity remained unchanged. If other wavelengths of light, e.g. only the UV content, were taken and used to irradiate the metal plate again, a change in speed was to be recorded. This situation could not be explained with classical physics. Einstein, too, had devoted himself to the problem in vain, until he came upon Max Planck's work.

A few years after Einstein was able to present the correct mathematical formulation for describing this phenomenon, which others had failed to do, he presented his light quantum hypothesis, describing light as a current of particles. These "quanta" (also called light quanta or photons), as Einstein pointed out, may occur only in certain portions and cannot be further decomposed (i.e. they can only be recorded or emitted as a whole). Imagine water from the tap: if you look at a stream of water in slow motion, the eye recognizes that it is composed of drops or small portions of particles.

Although there are still differing opinions as to the origin of the term "quanta," it is obvious that it goes back to Einstein.

Light according to Einstein's light quantum hypothesis:

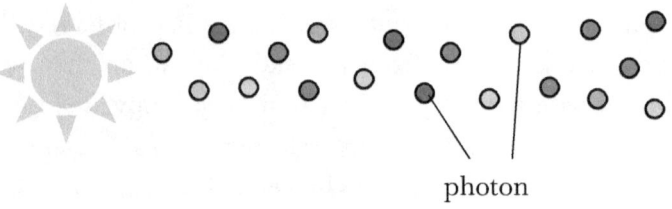

photon

Light according to classical physics:

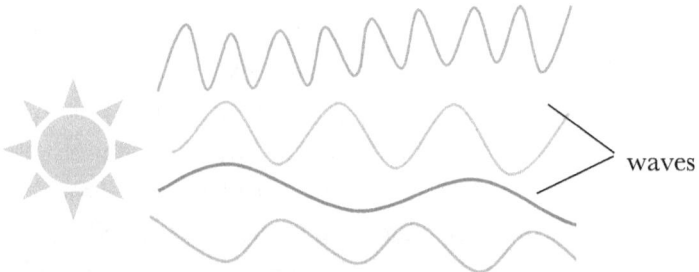

waves

Planck's formula, which had at first been regarded as an ordinary mathematical expression, was now corroborated by Einstein's work. Planck himself had probably underestimated its importance, or even regarded his own interpretations as incorrect, for only eight years later did he dare to describe energy states as discreet[9]. Even his previous, carefully formulated statement that radiation could be quantized gives insight into his uncertainty.

---

[9] Discreet processes seemingly contradict classical physics, which describes nature as a continuous process

By the time of the first Solvay Conference in 1911, his suspicion could be felt. Planck vehemently rejected the hypothesis that light is quantized. Instead, he noted that only energy exchange was quantized. Einstein's postulate must have shaken Planck's belief in classical physics so much that he presented a further theory that existed completely without quanta.

Max Planck was not alone in his mistrust of Einstein, for all the work of natural scientists, especially those who contradicted the general understanding at the time, had to be subjected to severe scrutiny before they garnered interest within scientific circles.

Planck was supported by American physicist Robert Millikan, who was determined to show that Einstein's work was a mistake. Millikan invested three years of his research into the photoelectric effect to refute Einstein's assumptions and he was taken aback when his research showed that Einstein's calculations precisely described the phenomenon. Ironically, among other things, this (counter) experiment won him a Nobel Prize in physics.

Planck and Einstein's discoveries were to be acknowledged after Millikan's contribution if they weren't already. But the dominant doubts and indecisiveness of scientists did not allow this immediately. The committee awarded them the Nobel Prize only years later (Planck 1920, Einstein 1921).

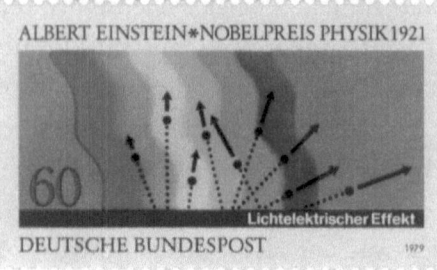

Special stamp for Einstein's
100th birthday

I want to spend the rest of my life
thinking about what light is.

Albert Einstein

Wave-Particle Dualism

Einstein and Planck's research results still had to contend with the acceptance of the masses. Thomas Young had experimentally proven the wave property of light a hundred years earlier with his double-slit experiment, and he was supported by Max von Laue's discovery in 1912, which showed the wave phenomena of X-rays on crystals. However, Albert Einstein postulated with his light quantum hypothesis that light is composed of small particles.

As is customary in the natural sciences, scientists returned here to classical experiments when hoping to obtain answers to new questions. So, in 1915, Young's double-split experiment was repeated again. Since Einstein's light quantum hypothesis that light consists of particles (photons), the question inevitably arose of which gap the photons would take on their way to the screen and how they would produce an interference pattern on the other side as a particle. For this purpose, the researchers modified the experiment in such a way as to answer these open questions.

At first, a weakly luminous light source was used, which produced only one photon when activated - like a single gunshot. The slits were each equipped with a detector, which was to trace the photon passing by.

The fact that the photodetector[10] functioned was possible thanks to the photo effect discovered by Einstein, in which light emitted electrons from the plate, the energy of which was converted into electric energy. When a photon flew through a slit, the activated detector shone. In addition, the usual screen was extended with a photo plate to mark the "impact point" of the photon with a black dot.

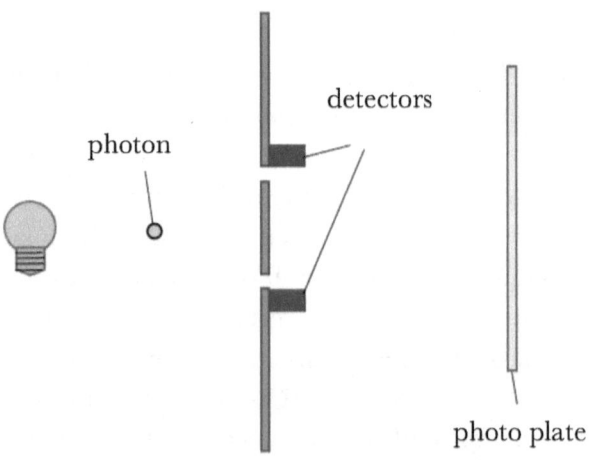

---

[10] Also called a light sensor or optic detector

With nothing else standing in its path, the experiment, which had already led to new insights into the nature of light, could begin. The photons were first passed through the splits while the detectors were inactive. The expectation was the following: the photon leaves the source, flies with a probability of fifty percent through the left or the right gap, and hits somewhere on the plate.

As expected, the photons passed the partition and landed on a random area of the photo plate - sometimes lower, sometimes higher, sometimes to the left, sometimes to the right, and sometimes in the center. At first there was nothing unusual about this, but as more and more photons left a mark on the plate and gradually formed a discernible pattern, the experimenters were astonished: the photo plate showed light interference.

Similar to Young's experiment, bright and dark spots had formed, which the photons had taken as a landing point or avoided. But how could that happen? The photons were shot individually and did not have the possibility to interfere with each other in front of the partition or afterwards. As if they were aware of one another, they dispersed themselves on the plate in the mold so that together they formed the interference pattern.

If the experiment was repeated in the same way, the marking points (or the landing points of the photons) were changed, but in general, they produced an interference pattern again, their distribution increasing in the middle and decreasing further from it. As in the case of wave phenomena, the highest intensity was focused in the center, where the wave crests or wave troughs were superimposed.

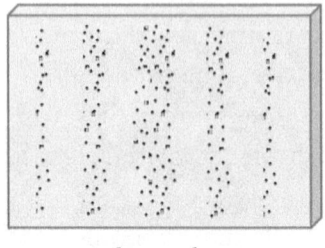 

photo plate				distribution of photons

At the beginning, the experimenters thought that the photon had somehow separated into two parts on the way to the plate. Only then would they be able to superimpose themselves and form an interference pattern. But in the next step of the experiment, they would learn something better.

In order to find out which gap the photons were flying through, the detectors were activated for the next experiment. Again the photons were fired separately, randomly taking sometimes the left and sometimes the right gap and landing on the photo plate. When a recognizable pattern was again to be seen, the new realization finally pulled the rug from under their feet. The photons had also avoided certain points, but this time they did not form an interference pattern, but rather two lines corresponding to the splits. During measurement by the detectors, they behaved like particles that passed through the gap on a direct path to the plate.

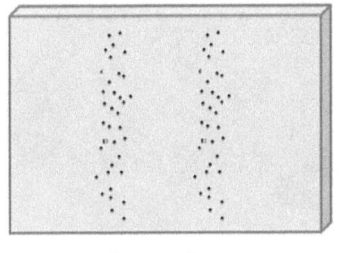
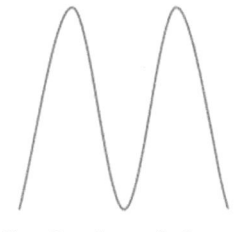

photo plate   distribution of photons

There had to be a clear explanation for this. Light could not behave as a wave and a particle at the same time. The experimenters desperately repeated the experiment many times in hopes of finding a meaningful explanation for this mysterious phenomenon. Always the same result: if the detectors were deactivated, the light behaved like a wave. If they were activated, it behaved like particles, and the interference pattern disappeared.

At the point of despair, science had no choice but to attribute a wave and particle property to light. With this step, so-called wave-particle dualism emerged, later to be dissolved by the Copenhagen interpretation.

Bohr's Atomic Model

The fact that the results of Albert Einstein and Max Planck's research, with their quantum character, were not immediately accepted did not surprise anyone at first. The time obviously was not yet ripe enough. But this did not prevent other physicists from applying their knowledge to their own work.

One of these was Danish physicist and later Nobel Laureate Niels Bohr. It was already known that each element had its own color spectrum. At the beginning of the 19th century, researchers had discovered fine, black streaks - so-called absorption lines - in the spectrum of sunlight. They initially searched in vain for explanations. It was not long before it was discovered that they were caused by cooler matter clouds in the solar atmosphere that absorbed certain frequencies. The light of other stars also had its own spectrum with its own absorption lines. Chemical investigations could conveniently be carried out from afar, e.g. determining what kinds of gases or elements made up the atmospheres of distant planets. It was only necessary to compare them with the spectral lines of the elements from Earth to know which elements were involved.

However, these phenomena can neither be explained nor described in a mathematically correct way using Rutherford's atomic model. It was time to revolutionize this outdated model. At this point, Niels Bohr's research on the light spectrum of hydrogen atoms would prove to be helpful.

The hydrogen atom is the simplest element and consists of one positively charged nucleus, the proton, and one negatively charged electron. Bohr tried to find a feasible description for this simplest of elements and to formulate his spectrum at the atomic level in order to apply this principle to all other elements.

First, he assumed that the electron emits a (specific) light only during the transition between the spectral lines. The electrons' energy levels could not possibly describe these jumps between the spectral lines. As Albert Einstein had managed to explain the photo effect through Max Planck's work, Niels Bohr also began to link other scientists' existing knowledge with his own research work.

Next, he integrated Planck's quantum of action into the existing atomic model and assumed that the atom's electrons only portion and cannot always emit or absorb one photon. It is worth mentioning that an electron emits a photon through its excess energy when it falls into a lower orbit and when it absorbs a photon as it ascends to a higher orbit. According to Bohr, therefore, the electron could only change orbits in leaps, which did not correspond to the classical image of continuity.

A further finding showed that these discontinuous jumps on larger electron paths were again classical. There was, obviously, a tangible transition from quantum mechanics to classical mechanics, which Bohr later formulated in his correspondence principle.

Having adopted Planck's quantum of action, he was able to describe the "quantum leaps" and the energy levels of the electrons in the hydrogen atom mathematically correctly. He received the Nobel Prize for this in 1922.

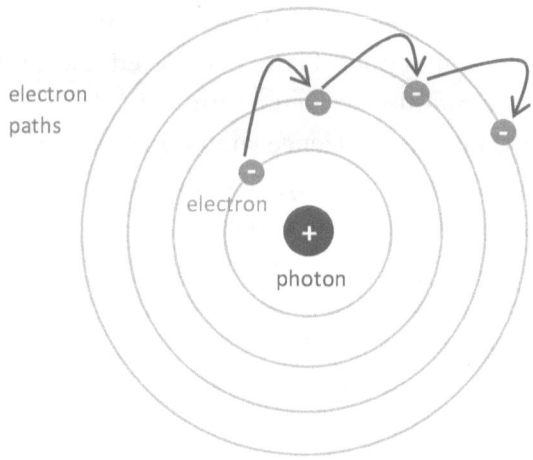

Atoms – The Invisible World

Today, Bohr's atomic model is the most common model depicted in most school materials. Its simple representation, which is similar to the familiar solar system, is easy to understand, but it is of course only an abstraction to give us a picture of atomic processes on a mathematical level.

This abstract image, which forms the foundation of all natural processes, has the task of bringing us closer to reality and its laws. Admittedly, it was even more paradoxical for many people at the time that our idea of nature was based on something which cannot be observed with either the naked eye or technical tools.

The Greek philosopher Democritus of Abdera and his teacher Leucippus had described the composition of all of nature, which was made up of the smallest indivisible units, so-called atoms. Since then, a dozen new models with similar postulates have emerged. It is interesting, however, that the ancient philosopher Plato considered the atom to be divisible, in opposition to Democritus.

The current state of science shows us that even the nucleus and the atom's circulating electrons can be broken down into further parts. We can observe these decompositions at CERN using computer simulations and measure their charges with sensitive sensors, even if other physicists like Werner Heisenberg and Hans-Peter Dürr object to the decomposition process and instead interpret the result of the collision of atomic particles as the creation of new elementary particles of the same group.

The atomic dimension is and remains much too small for us to ever observe, even using optical instruments like the scanning tunneling microscope (STM), a high-end device whose tip ideally is composed of only one atom and that can only appease us with measured values.

As early as 1871, Austrian philosopher and scientist Ernst Mach rejected the existence of atoms vehemently, for the reason that they are not visible or perceptible with the senses. But science will always be accompanied by such critical voices, and this is also necessary for its progress.

Bohr's atomic model was no exception here, either, and it had to stand up to great criticism. One of the leading critics was undoubtedly Austrian physicist and later Nobel Laureate Erwin Schrödinger, who compared Niels Bohr's jumping electrons to fleas.

The very original idea of a French physicist would later give Schrödinger food for thought resulting in his world-famous wave equation. Schrödinger's attempt to break down Bohr's atom with his subsequent wave mechanics would inevitably lead to a difficult relationship between him and Bohr, but today we know that Bohr's atomic theory, which once postulated "jumps" of the electrons in an atom through the inclusion of quantum theory, is obsolete. These electron paths do not exist at all, which will be discussed in more detail below.

Niels Bohr

The Matter Wave

In 1923, almost a decade after Einstein's photo effect, the light quantum hypothesis gradually gained acceptance. This was mainly due to the discovery of American physicist and later Nobel Prize Laureate (1927) Arthur Compton.

Compton had long studied X-rays and was able to explain the interaction between electrons and X-rays with Einstein's light quanta (photons). Light consisted of a stream of particles - but at the same time it also had wave properties. Both theories of light seemed to have been demonstrated through previous experiments. As in the case of electromagnetism, light was evidently two sides of the same coin. A logical theory that could combine both qualities into a common existence was missing, though.

Scientists were hoping to find the solution in quantum physics, which was further investigated and further developed and was later to replace wave-particle dualism. While many scientists were pursuing their duties, the extraordinary theory of a French physicist showed that this dual nature included more than just light.

De Broglie's Bold Idea

It has hitherto been assumed that matter is composed of small particles like molecules and atoms. This could also be explained logically, since its macroscopic nature can be registered with the sensory organs, e.g. by touching or seeing. In addition to this, matter, unlike light, has a rest mass. Could light consist of particles, since it is not tangible and has no rest mass? Or can it be weighed in some form?

The ordinary human understanding was programmed to accept matter as particles and light as a wave. It is not surprising that the particle characterization of light was not initially taken into account, and even theories in this direction were rejected. On the other hand, nobody ever thought about the wave property of electrons and other atomic components. That is, until a French aristocrat and later Nobel Laureate presented a bold theory that turned physics upside down.

Bohr's atomic model, with electrons that jumped from shell to shell like fleas, allowed the spectra of elements to be correctly described, but it did not provide a rational and physical explanation. Above all, the correlation between Bohr electrons and Planck's quantum of action was missing, not to mention the fact that his model could not answer to every phenomenon and partly led to contradictions.

Louis de Broglie had been dealing with the dual nature of light since the beginning of the 1920s. In his experiments, he noted how light showed its wave characteristics on long time intervals. But on brief snapshots, for example, in energy exchanges between light and matter, it behaved like a particle. His conclusion from this observation was that the property of light changed according to the situation, but both were never true at the same time.

De Broglie sought a plausible explanation. If light could be dualistic, what then spoke against a dual nature in electrons? He had probably considered this in that way when he thought about the wave properties of matter.

He found the simple explanation in violin strings. He imagined the electron paths like oscillating strings, which are fastened to their two ends and have their own wavelengths. As Einstein and Bohr once did, De Broglie also used Planck's quantum of action for his research and determined the wavelength of the vibrating strings. These strings are referred to as "standing waves," which means that they vibrate up and down without moving along the string.

$$\lambda = \frac{h}{p}$$

p... impulse
h... Planck's quantum of action

If Bohr's atom could be like a miniature model of the solar system and its electrons jumped like fleas, as Schrödinger had critically identified, why couldn't it also be a tiny violin with strings?

Standing wave:

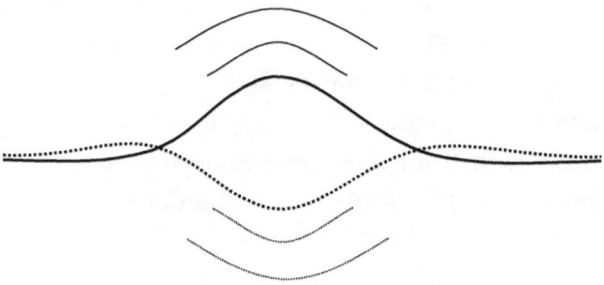

Atomic model according to De Broglie:

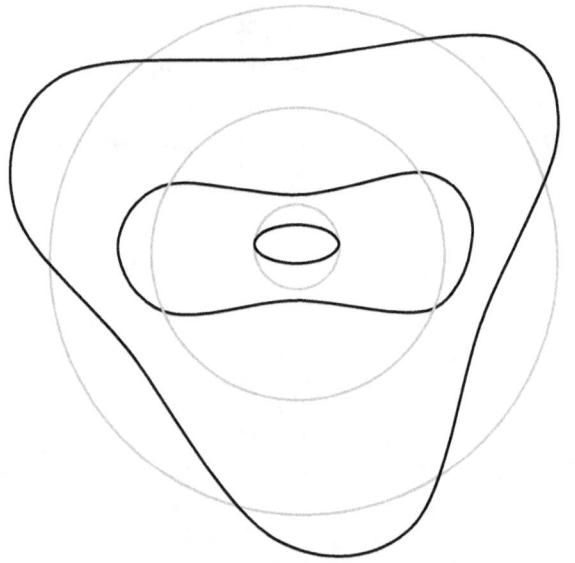

The outer shell consists of six standing waves, the middle of four and the inner of two.

On the basis of these unusual ideas, De Broglie wrote his work and then published it as a doctoral thesis. The problem with the whole thing was that it was merely a theory. It really did not have experimental support. Imagine that atoms were a pot of gold surrounded by goblins. This fantasy could be formulated mathematically, but to experimentally prove it would be very difficult. So finally the perplexed investigators turned to Albert Einstein, who commented on De Broglie's theory with the words, "It may look crazy, but it is really logical!"

Einstein turned out to be right. A short time later, Clinton Davisson was finally able to explain the previously unresolved pattern of his experiment. The electrons that Davisson had experimented with had produced a wave-like reflection pattern. But now he finally had an explanation thanks to De Broglie's idea.

Six years later, Louis de Broglie received the Nobel Prize for his original work and in 1937 Davisson received it for experimental confirmation.

Louis de Broglie

Schrödinger's Wave Equation

Not even months had passed since the publication of his work, and there were already physicists pouncing on De

Broglie's electron waves. De Broglie's bold idea had given science a further breakthrough, but it could not justify its existence in a working atomic model. There was also the missing mathematical formulation to describe the energy exchange of material waves and the emission and absorption of light.

First attempts came from Austrian physicist and later Nobel Prize Laureate (1933) Erwin Schrödinger. He first failed with his waveforms, which incompletely described the light spectrum of the hydrogen atom. This incompleteness was not due to his intellectual creative power, but to the state of knowledge of physics.

In the same year, the German physicists Werner Heisenberg, Max Born, and Pascual Jordan worked on their own quantum mechanics design for the atomic model, which they termed matrix mechanics. They also encountered the same problem as Schrödinger, but thanks to the "spin"[11] introduced by Wolfgang Pauli, they could provide mathematical evidence. Schrödinger didn't know about spin and thus could not provide a solution, so he revised his approach. One year later, when he published his final work with a new foundation, he came into the limelight as the savior of quantum theory.

The general form of Schrödinger's equation:

$$i\hbar \frac{\partial}{\partial t} | \psi(t) \rangle = \hat{H} | \psi(t) \rangle$$

---

[11] Spin: angular momentum of electrons

The general form of matrix mechanics:

$$i\hbar \frac{\partial}{\partial t} \langle \phi_n | \psi(t) \rangle = \sum_m \langle \phi_m | H | \phi_m \rangle \langle \phi_m | \psi(t) \rangle$$

However different the formulas looked at the first glance, it turned out a little later that the two were equivalent. Both formulas ultimately led to the same results, but their creators had taken different routes.

However, Schrödinger's wave equation had a significant advantage: Schrödinger had relied on mathematical elements from the known fields of physics when building his wave formula. They were equally conventional in optics, acoustics, hydrodynamics, etc. This is why his equation received a higher degree of acceptance: it was more familiar among physicists. Albert Einstein also characterized Schrödinger's work as brilliance and similar laudatory feedback came from Max Planck, who was said to have read Schrödinger's work like a curious child. The wave equation revived physicists' hopes that quantum mechanics was compatible with the classical worldview. That meant the end of Bohr's jumping electrons.

Niels Bohr did not seem to be delighted by these pioneering developments. In the same year that Schrödinger's work was published, Bohr invited Schrödinger to Copenhagen. An understanding was to be made between his equation and matrix mechanics. According to a report by Werner Heisenberg, who was also at the Bohr Institute, the discussions between Bohr and Schrödinger were very intense. Schrödinger lived in the same house as Bohr and their controversial discussions went on from early in the morning to late at night.

Bohr was said to be very uncooperative, according to Heisenberg, almost "like a relentless fanatic," so Schrödinger, exhausted, fell ill with a fever. Schrödinger had to stay in bed for days, while Bohr sat on the edge of the bed and continued trying to convince him. Bohr's negative attitude was not quite unfounded. Schrödinger's wave equation also had its drawbacks. It failed with "multi-electron problems" and also had difficulties with the transition from the microcosm to the macrocosm. According to his wave equation, there would no longer be any stable objects. Schrödinger believed that it was only a matter of time, though, before these gaps would be filled mathematically.

Erwin Schrödinger

The Uncertainty of the Electron

Schrödinger's wave equation was able to answer to a large part of phenomena. However, it concealed its own specific problems. In addition, further questions emerged, for example, how to imagine the waves in an atom concretely, but above all else: What was to happen with the particle image, which could also allow a description of the phenomena?

Max Born, a German physicist and later Nobel Laureate (1954), was concerned with these topical questions of quantum theory, like many of his colleagues. Born did not start out stubbornly assuming a wave as Schrödinger did, and instead, he also included the particle property of the electron. Both were to represent a complementary whole, similar to the dual nature of light, so somewhere, there should be a particle in water that triggers waves.

In accordance with his impression, Born conducted experiments with an electron beam to capture the position of a single electron. When he took advantage of Schrödinger's equation, he found that the electron particles were directed by waves, and their wave strength determined an electron's probability of striking, similar to a ball floating to shore on a wave.

Thus, Schrödinger's wave equation was nothing more than a probability function applied in statistics to calculate probabilities. However, this fact was only shown in experiments with a high concentration of particles, as with the electron beam. Therefore, it was time for a new, contradictory interpretation, which was later to divide natural scientists.

Max Born

Heisenberg's Uncertainty Principle

Max Born's postulated probabilistic function formed a further ring in the long discourse. On that basis, German physicist and later Nobel Prize Laureate (1932) Werner Heisenberg worked on a mathematical formula that was to determine the movement impulse and the location of an atomic particle. Heisenberg had been inspired by ancient philosophy since his youth. He had used Plato and Aristotle's ideas as a source of reflection, as he stated in his memoirs, and for exciting conversations with his hiking friends, he talked about such things as Democritus' atom and Plato's parables. Inspired by Plato's cave allegory, the young scientist had made his way to the hidden reality. Heisenberg succeeded in brightening their shadows. However, he feared that no one would understand him - and his fears would later come true.

In his intensive work on developing a mathematical formula to determine the momentum and location of an atomic particle, he surprisingly found that both are not possible at the same time. If the momentum were to be determined more precisely, the particle was no longer to be found; it became "unfocused." If, on the other hand, the location of the particle was to be found, the momentum could not be determined. The electron could, as it were, be found everywhere in the atom, and it was only through measurement (disruption) that it appeared and disappeared again in naught. The possibility of tracking the electron in its "orbit" was impossible because of the missing momentum data. The difficulty lay not in technical inadequacies, but in the natural laws of the microcosm.

This realization completely contradicted the classical worldview. For example, with a radar gun, both the speed and the location of a vehicle can be precisely measured, thereby determining its future location. But in the world of the microscopic, the location of a particle could only be expressed in probabilities. Thus, Heisenberg concluded that the path of the electron only comes to be through measurement or observation. As long as the elementary particles are not measured, they would be in a certain floating state.

This postulate, which Heisenberg described as uncertainty or indeterminacy principle, again shook the world of classical physics.

Heisenberg on a German stamp

Zeno's Arrow Paradox

It is nowhere recorded whether Heisenberg was inspired by Zeno's Arrow Paradox in his discovery of the uncertainty principle. It is indisputable that the ancient, more than two-thousand-year-old paradox describes a similar problem.

The Greek philosopher Zeno of Elea had already taken an interest in space, time, and movement in Antiquity. He described the relationship between these parameters by means of paradoxical examples. The closest to Heisenberg's uncertainty principle is definitely the arrow paradox. Perhaps at first glance it may seem unusual that, long before Heisenberg came to be aware of it, Greek philosophers had already occupied themselves with more or less the same thing, without technical or mathematical tools. But Zeno's Arrow Paradox illustrates how important philosophical approaches are through the difficulty of determining the location and velocity of an arrow as a localizing parameter.

Suppose we shoot an arrow with a bow and then watch it as it flies from point A to point B. How can the exact location of a moving arrow be determined? Quite simply, we freeze the picture and see where the arrow is. But what about its movement? According to Zeno, its movement simply does not exist if we want to capture the position of the arrow. This consideration on Zeno's part is sensational in view of its time, considering that not even the anatomy or function of the eye had been explored.

The human eye captures about 24 frames per second - like a camera that takes several images in sequence. If, within one second, the arrow travels a certain distance from point A to point B, we basically see 24 different positions of the arrow during this distance. The addition of these momentary impressions then results in the movement, which we perceive as a fluid sequence. A fly, for example, takes 600 frames per second. That's why movement is slower for them. If our eye were capable of capturing an infinite number of images, the arrow would never move an inch, let alone from point A to point B.

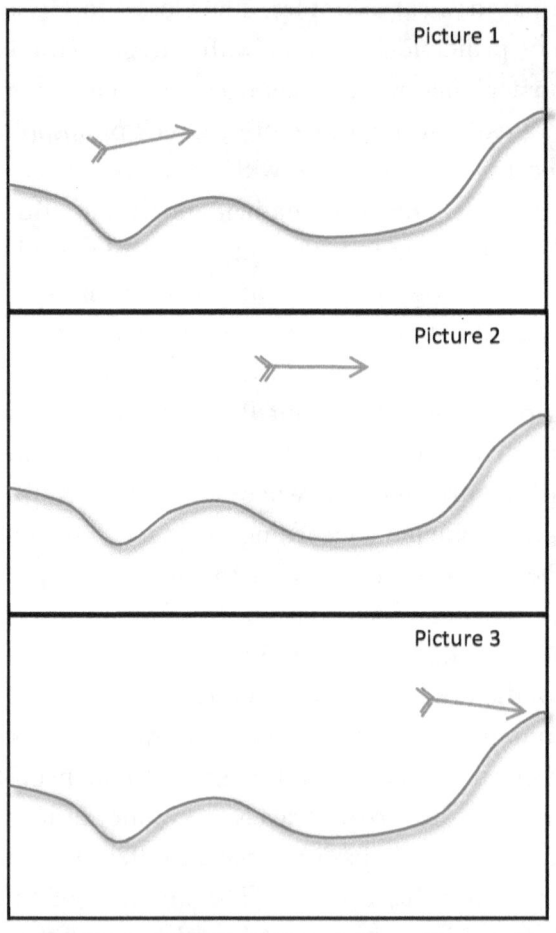

The addition of the pictures as movement perception:

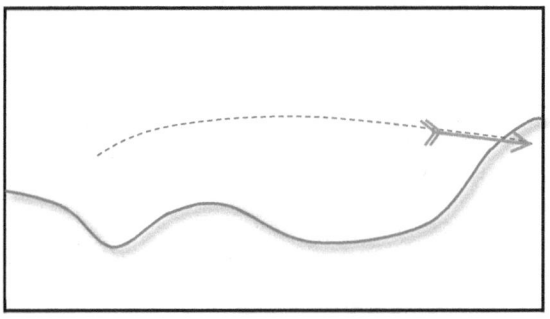

Zeno of Elea

The Copenhagen Interpretation

Schrödinger's wave equation, even in its imperfect form, had been a great (intermediate) step in quantum theory. Max Born had become acquainted with it and summarized the equation as a so-called probability function, which is also used in statistics. With Heisenberg's uncertainty principle, Born's postulate became an additional support. The electron was not a wave, which Schrödinger continued to insist on, but a particle as long as it was measured. Without measurement, it was everywhere - similar to the double-slit experiment. The wave and interference were picked up through measurement. The subject, that is, the observer, was clearly part of the system, which he "disturbed" through measurement.

It was not long before we realized that this fundamental news required a new interpretation for quantum theory. This was to happen in Bohr's homeland, in Copenhagen. Bohr had looked at Born and Heisenberg' research, and he was able to agree with them more than with Schrödinger or De Broglie's electron wave.

Bohr continued to believe that electrons are particles, but he did not refuse their simultaneous wave character. On the contrary, he underlined the duality of the electron and of the photon. In his complementary principle, he described the fact that electrons possess two properties, but that only one occurs according to the experiment. He strictly rejected that the electron could only be a wave, as postulated by Schrödinger.

On the basis of these new findings, in particular Max Born's probability function and Heisenberg's uncertainty principle, Bohr, together with Werner Heisenberg, finally formulated a conflict-free interpretation that went down in history as the "Copenhagen interpretation." According to this, predictions of natural processes at the atomic level are not possible, but can only be expressed in probabilities. The probability function collapses through measurement and the particle appears. The core statement is that the foundation of our reality is based on coincidence and only comes through observation, just as the electron only appears through measurement. The Copenhagen interpretation still represents the status quo of quantum physics today.

The 5th Solvay Conference

In the autumn of that same year that Bohr and Heisenberg founded a conflict-free interpretation of quantum theory, the Fifth Solvay Conference took place. It was among the most famous and exciting conferences in history, with great personalities like Einstein, Bohr, Schrödinger, Dirac, Born, Heisenberg, Pauli, and other physicists. The focus of the conference was "Electrons And Photons," as well as their interpretations, which was to prevail despite criticism.

Schrödinger, who had not spoken to Bohr critically about his atomic model, was shaken when he saw how his famous wave equation had been interpreted, for he still represented the wave hypothesis that Bohr decisively rejected. The ancestors of quantum theory, like Albert Einstein and Max Planck, reacted just as critically when they caught wind of the probability interpretation, which was too much even for Einstein. For this reason, he asked cynically whether the moon existed if one did not look, as an allusion to the collapsing wave functions. "God doesn't play odds," he added, with regard to the postulated likelihoods. The world cannot consist of coincidences - there must be hidden variables that have not yet been discovered, according to Einstein. The Copenhagen interpretation not only contradicted one part of classical physics, as in the discussion of light, but finally left its foundation shattered.

Discussion about this interpretation continued to be controversial. The exchange of blows was particularly violent between Bohr and Einstein when they tried to convince each other with their thought experiments. However, Bohr always managed to disprove his adversary's thoughts, some even by referring to Einstein's theory of relativity. He hit him, so to speak, with his own weapons.

In the end, the majority of physicists advocated Bohr and Heisenberg's conflict-free interpretation, while a smaller group consisting of Albert Einstein, Max Planck, Erwin Schrödinger, and Max von Laue postulated the Copenhagen interpretation, in vain, as an incomplete interpretation of quantum theory. The Fifth Solvay Conference ended in favor of Bohr and his entourage.

The Fifth Solvay Conference on "Electrons And Photons"

The End of a Quantum Section

The first thirty years of the twentieth century thus had many a fascinating discovery. The contributions of scientists had shaped atom physics in such a way that they hardly corresponded to the classical physics of earlier times. Max Planck's professor, who had told him that everything had already been discovered in physics, had been wrong. Nature had turned out to be a barrel full of surprises. Their laws seemed to behave quite differently in the microcosm than in the macrocosm.

With the Fifth Solvay Conference, quantum theory had now obtained its conflict-free interpretation. But there was still no common, functional theory that covered both worlds. But until then, all mathematical experiments had been in vain. The foundation of reality actually seemed to be based on probabilities. So did God play odds?

Not all physicists agreed with the current statistical interpretation. Therefore, the next section will show the "attacks" against the Copenhagen interpretation and alternative interpretations of quantum theory. But first, two personalities have to be briefly mentioned, who, like their colleagues, made an important contribution to quantum theory and should not be forgotten.

Paul Dirac, British physicist and Nobel Laureate (1933), is one of the co-founders of quantum physics. In 1928, he succeeded in mathematically combining Einstein's special relativity theory with quantum theory. Independently of Schrödinger, he proved the equivalence of his wave equation and Heisenberg's matrix mechanics. In addition, he came up with the idea of  path integrals, which showed all possible paths a particle could take from point A to point B. It was thanks to Richard Feynman that the idea was taken seriously for the first time. Moreover, he introduced the Dirac equation, named after him, and developed many other interesting theories on a mathematical basis.

Wolfgang Pauli, Austrian physicist and Nobel Laureate (1945), is also one of the co-founders of quantum physics. Important contributions to modern physics and innumerable publications (93 articles and 11 books) bear his manuscript. His most famous work is the "Pauli Exclusion Principle." He explains the structure of the atom in which no two elec-  trons can agree in all (four) quantum numbers and mutually exclude each other, so to speak.

Alternatives to the Copenhagen Interpretation

It would be careless to assume that the Copenhagen interpretation would not face differentiated views and critical remarks. Just as all previous questions in science had led to controversy, including the "indivisible" atom or "infinite" light, the Copenhagen interpretation was also questioned or even modified by small circles.

It is the nature of a scientist to question even the foundations of the existing systems, which have earned their legitimacy. Today there are more than a dozen alternative interpretations of quantum theory. Some try to restore a bridge with classical physics - others are even crazier than quantum theory. However, I would like to limit myself to the most popular examples.

The EPR Paradox

Despite losing the fight against Bohr's reasoning, Einstein was still convinced that this was an incomplete interpretation of quantum theory.

About eight years had elapsed since the Fifth Solvay Conference, and Einstein continued to seek support for his postulate, while most physicists adopted quantum physics as it now was, dealing with mathematical subtleties rather than basic things.

For some, it may have seemed that Einstein had not understood quantum theory, but he understood it perfectly. He could not reconcile its indeterministic character, the idea that everything was to be based on chance, with his worldview, just as Max Planck could not reconcile the light quantum hypothesis with his classical worldview. Just as other scientists had once tried to disprove Einstein's light quantum hypothesis and photo effect, now Einstein tried to expose the Copenhagen interpretation. Together with his colleagues Podolsky and Rosen (together called EPR), they published a thought experiment intended to demonstrate the incompleteness of quantum theory with a contradiction.

They first showed that either the position or momentum of a particle is predictable, therefore not contradicting Heisenberg's uncertainty principle. They further argued that the choice of both properties depends not on the particle, but on the decision of the observer. It means the observer disturbs the particle in such a way that only one property remains and the other disappears as if it had never existed. But somehow the particle would have to exist, for how could it simply disappear from the scene? There must be an objective reality, even if no one is looking.

As a solution, Einstein, Podolsky, and Rosen considered the following: The observer takes the measurement indirectly over another particle, so that it does not disturb the target object. Let us now imagine two particles that interact with one another and then fly apart. Since they are no longer juxtaposed, they cannot correlate with one another. They move apart very quickly, which is why we can detect the momentum of one particle and at the same time the location of the second particle. This means that the location and impulse would thus be determined at the same time, but this would be a contradiction to the uncertainty, which is the foundation of the Copenhagen interpretation.

The Copenhagen supporters had the choice: either a gap in the Copenhagen interpretation, which would mean that it was incomplete and that "hidden variables" exist, or that there is a "spooky connection" between the particles that "communicate" with each other and cause the partner particle to be "blurred." This would mean that reality is not local. This, in turn, would result from the fact that there could be no "free space" at all, but that everything is connected with everything.

Einstein at first seemed to have pushed his opponents into a corner with his sharp-witted thought experiment. According to his theory of relativity, a "spooky connection" with the photons would not be conceivable since nothing could move faster than light - just as unthinkable would be the non-local reality, which would, however, be applicable in that case. The Copenhagen community could understand the logic behind it, but they argued against Einstein again because the indirect measurement on the second particle is not a proper measurement.

Thirty years later, after Einstein's death, the EPR paradox was finally to be disproved experimentally. To the great surprise of all involved, Einstein had been right with the "spooky connection" in favor of his opponents, and it actually existed. The particles did not communicate with each other, but were associated with the same, non-local system.

Schrödinger's Dead-Living Cat

Erwin Schrödinger was known for formulating complex facts with simple explanations. Ignoring the conclusions of the Fifth Solvay Conference, he also worked for several years to make an absurdity of the Copenhagen interpretation, even though the new interpretation of quantum theory used a modified form of his wave equation.

In the same year (1935), Schrödinger published a simple thought experiment that, in comparison to Einstein's EPR paradox, has not lost its popularity to this day. Imagine a closed box with a cat and a mechanism consisting of an unstable nucleus and a vial of poison. We know that the decay of an atomic nucleus is impossible to predict. Eventually it will decay and release the deadly poison from the vial. When that happens, the cat is dead.

We can only ascertain if this decay has taken place and killed the cat by opening the crate and looking. But how can we define the state of the cat when we have closed it in the box and do not know if it is still alive? According to quantum mechanics, it is in a state of so-called superposition where its states become superimposed. This means that the cat is dead and alive, so long as we do not look into the box or take a measurement.

If we now open the crate and look inside, the wave function collapses and with it the hovering state. Then we can see what has happened to the cat. If we are lucky, it is alive; if not, then we buy a new one at the next pet shop and hopefully won't kill it.

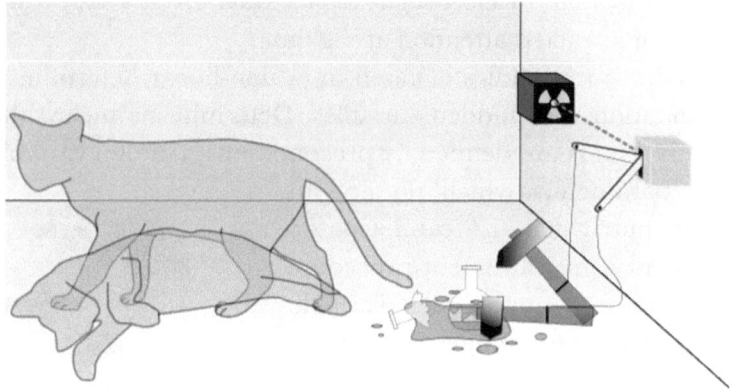

But all joking aside, how is this supposed to happen? How can the cat be alive and dead at the same time? This is a contradiction in and of itself. However, the thought experiment could correspond to the Copenhagen interpretation, if it were applied to macroscopic objects.

Many people misunderstand Schrödinger's cat as a simple explanation of quantum theory. But Schrödinger intended to critically illustrate the "crazy" side of the Copenhagen interpretation and the basis for our reality. However, this "insanity" was to receive its experimental confirmation in the 21st century.

De Broglie-Bohm Theory

The American physicist and philosopher David Bohm represented the Copenhagen interpretation until the end of the 1940s. Later, he began to question it. As a logical consequence, he introduced an alternate interpretation, so-called Bohm mechanics. He did not realize until much later that it was equivalent to De Broglie's pilot wave from 1920, which had not attracted attention at the time.

Bohm's mechanics is based on a non-linear deterministic foundation with hidden variables. Deterministic means that future events are defined by preconditions. Hidden variables are parameters which do not appear, but which also cannot be derived in a measurement procedure. Determinism and hidden variables represent the counterpart of the chance principle in quantum theory. With his theory, Bohm was able to reproduce predictions like the Copenhagen interpretation. Only the postulate of disturbance through measurement or observation was strictly rejected by Bohm.

David Bohm

Bell's Inequality

Since the Copenhagen interpretation, one of the greatest controversies of the current quantum theory was non-locality. For this reason, its opponents, like Einstein and David Bohm, resorted to the postulate of hidden variables to justify this incompleteness. However, no solid evidence could be provided for or against this postulate, until John Bell, a Northern Irish physicist, began to deal with the EPR paradox and Bohm's theory.

Bell had dealt both philosophically and mathematically with quantum theories, especially with the antitheses. He finally published a mathematical paper on this approach, known as "Bell Theorem" or "Bell's Inequality," which shed light on the matter in 1964. The core statement of his inequality was that there could be no theory of local hidden variables reproducing the statistical predictions of quantum theory. According to Bell's Theorem, the theories of Einstein, Bohm, and of all others who advocate hidden variables fulfilled the inequality and quit. The values calculated with quantum theory according to the Copenhagen interpretation, on the other hand, violated Bell's inequality - violated in this case means that they contradicted the assumption of the locality and possibility of hidden variables. Quantum theory according to the Copenhagen interpretation is thus not local, but "complete," contrary to the postulates of his opponents.

Many-Worlds Interpretation

One bizarre theory is certainly the "many-worlds interpretation" by American physicist Hugh Everett. However, it is not an alternative to the current quantum theories, but rather, it attempts to describe the collapse of the wave function according to the Copenhagen interpretation by means of worlds. The collapsing wave function had led to controversial discussions, but with Everett's explanation, they were to reach a new level that made the issue even more controversial.

At first, Everett abandoned the collapse postulate and tried to describe the measuring process using Schrödinger's equation. In doing so, he did not define the observer as an interferer, but as a person whose state is altered through measurement (i.e. the reverse of the Copenhagen interpretation). This theory applied to Schrödinger's thought experiment provides the following result: the cat in the crate, fighting with death, satisfies two conditions simultaneously. At the moment the observer opens the box, he sees the cat only in a certain state, either alive or dead, so each state has a 50% probability. However, according to Everett, the unsuccessful state does not simply dissipate in the air. It already existed before, so it must continue to exist, not just here, but in another world. At the moment the observer opens the crate, he lays hold of the cat in one state. He sees the other state in another world, according to many-world theory.

The world is divided into other worlds.

Thus, the hovering postulated by the Copenhagen interpretation becomes obsolete. The cat has always had both states and they both prove true, without anything collapsing.

The same theory applied to the double slit means that the universe divides into two worlds where the photon flies through the left slit in one world and through the right slit in another. The interaction of the two worlds then yields the interference pattern. The question is, then, how can the process with activated detectors be explained, whereby the interference pattern is extinguished? There seems to be no answer. Even Everett's theory can only interpret one part and will remain a theory.

However, it must be emphasized that Everett did not imagine the worlds as being spatially separate, as parallel worlds or time travel like in science fiction films, but as separate states in the respective state space. Nevertheless, there was no way to prevent this bizarre idea of parallel worlds from being envisioned.

One of the most famous supporters of Everett's theory is Israeli-British physicist David Deutsch, who is mainly concerned with parallel universes based on the many-worlds theory.

Hugh Everett

Experimental Evidence

As a rule, there is theory first, followed by experimental evidence that confirms or rejects it. Whether Max Planck's quantum theories, the theory of relativity, the photo effect, the light quantum hypothesis, or De Broglie's matter-wave, they all had to undergo rigorous examination.

With new insights, scientists continually revert to classical experiments and observe the new behavior of the experimental regimes in order to conclude new results. This approach was to continue proving itself in modern physics. That is why, in this chapter, I will present some interesting evidence that I consider necessary.

Double-Slit Experiment with Atomic Particles

At the beginning of the nineteenth century, Thomas Young had proven the dual nature of the double-slit experiment. In 1915 physicists repeated it with photons due to new insights and the dual-nature of light. In the 1920s, Louis de Broglie postulated the electron wave, which had been detected using X-rays. Then the German physicist Claus Jönsson carried out the double-slit experiment with electrons in 1961 and behold: they actually behave like photons, sometimes as waves and sometimes as particles, of course depending on measurement (detectors).

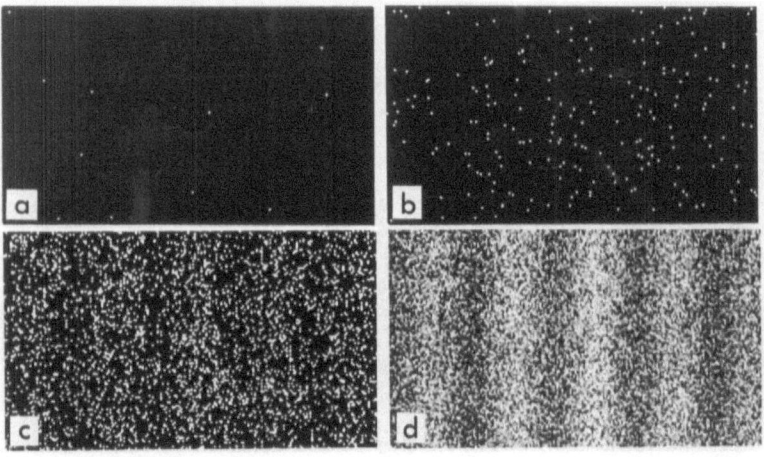

(a) 11 electrons
(b) 200 electrons
(c) 6,000 electrons
(d) 140,000 electrons

The double-slit experiment was carried out over time with much larger particles such as atoms and fullerenes. Fullerenes consist of sixty carbon atoms and are constructed like a soccer ball. In contrast to photons, they are not only considerably larger but also have masses. However, the wave pattern of fullerenes only shows up in the vacuum. In a normal environment, they interact with equal-sized air molecules, which inadvertently results in constant local measurements. As a result, no interference appears. This also explains why in everyday life we have not heard of quantum phenomena. Therefore, the moon also exists when no one is looking.

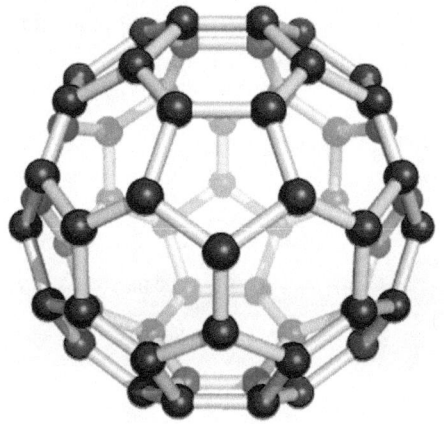

The structure of fullerenes

Quantum Entanglement

Another of the strange phenomena of quantum physics is entanglement. In his EPR paradox, Einstein had ruled out a "spooky action at a distance" between the particles because it was contrary to classical physics. However, a group of scientists succeeded in observing the action at a distance, also called entanglement.

First, imagine two particles that are not spatially separated from each other. For example, when a photon is transmitted by a crystal, it is divided into two pairs. One of the photons is guided in one direction, while the second is guided in the opposite direction, which means they are spatially separated from each other. If detectors are placed along their path to measure their properties, it can astonishingly be seen that they always act the same. One property, for example, can be polarization, i.e. the photon's orientation.

Possible polarizations:

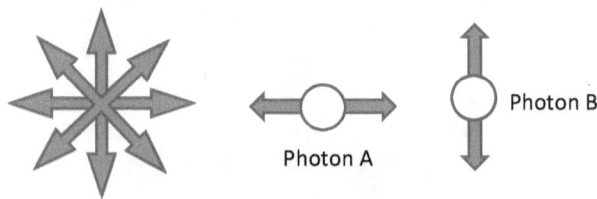

If photon A with a horizontal polarization is measured, then the measurement on the second photon B shows a vertical polarization. If the polarization of a photon is "changed" by so-called polarization filters, the polarization of the second photon also changes. No matter how many times the experiment is repeated, the photons behave in a mutually interrelated manner.

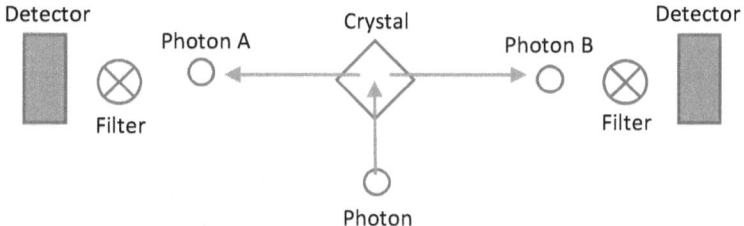

Einstein's "spooky action at a distance" seems to exist. According to Bell's inequality, this experimental evidence was a further confirmation of the non-locality of the microcosm, which Einstein could no longer witness. Today, research on the phenomenon of quantum entanglement is being pursued for use in cryptography. Within the framework of this research, new insights are constantly being gained. Recently, researchers found that the action at a distance between two particles is ten thousand times faster than light. This clearly disproves Einstein's thesis, which postulated the speed of light as the maximum.

A Quantum Mouse in Schrödinger's Crate

Two French physicists and Nobel Prize Laureates (2012), Serge Haroche and Jean-Michel Raimond, succeeded in creating an experiment in 1996 that allowed them to penetrate the deep world of Schrödinger's crate. They first put the state of a rubidium atom into superposition by means of lasers and then sent it into a hollow space. It almost represented Schrodinger's dead-living cat. Then they let another atom, called the quantum mouse, fly into the space. This atom had the task of observing the "cat" without opening the "crate." In that way, Haroche and Raimond were able to show that the transition from the microcosm to the macrocosm occurs gradually and that superimpositions collapse more rapidly as the system grows.

A Quantum Mouse in Schrödinger's Crate

Epilogue

The paradoxical character of quantum physics can no longer be denied now that you have finished reading this book. In the beginning, quantum physics may seem almost absurd to an outsider. I at least felt this way when I had to give a presentation on quantum cryptography during my university studies. Among so many different cryptographic methods, your humble author picked none other than this topic – what's more, in English, my third language! But thanks to the quanta, everything went well.

Looking back, I see that the difficulties in understanding quantum physics do not lie in its paradoxical realities. Rather, I see the hurdle in its scope. It is only necessary to imagine that every theoretical, mathematical, and experimental approach summarized in this book is a matter of years of work. And surely there are a number of other facts that this introductory book has not even touch on, for good reason.

As already stated in the preface, this reading should limit the complex subject to the essentials. For deeper insight, I have included a bibliography after this epilogue.

Bibliography

Fred Alan Wolf: Der Quantensprung ist kein Geheimnis

John Gribbin: Auf der Suche nach Schrödingers Katze

Anton Zeilinger: Einsteins Schleier

Brigitte Rothlein: Schrödingers Katze

Richard P. Feynman: QED – Die seltsame Theorie des Lichts und der Materie

Hans Lüth: Quantenphysik in der Nanowelt

Herbert Pietschmann: Quantenmechanik verstehen

Werner Heisenberg: Quantenmechanik & Philosophie

Further recommended literature

Werner Martienssen & Dieter Ross: Physik im 21. Jahrhundert

Werner Heisenberg: Physik der Atomkerne

Werner Heisenberg: Der Teil und das Ganze

Erwin Schrödinger: Was ist Leben?

Dieter Hoffmann: Erwin Schrödinger - Biographien hervorragender Naturwissenschaftler, Techniker und Mediziner

Jukka Maalampi: Die Weltlinie – Albert Einstein und die moderne Physik

Picture Credits

p. 32: "Niederdruckmetalldampflampe" from Sheevar – own work. Licensed under Creative Commons Attribution-Share Alike 3.0 via Wikimedia Commons - http://commons.wikimedia.org/wiki/File:Niederdruckmetalldampfla mpe.jpg#mediaviewer/File:Niederdruckmetalldampflampe.jpg

p. 79: "Schrodinger's cat" from Dhatfield – own work. Licensed under CC BY-SA 3.0 via Wikimedia Commons - http://commons.wikimedia.org/wiki/File:Schrodingers_cat.svg#me diaviewer/File:Schrodingers_cat.svg

p. 80: "David Bohm" from Original uploader was Karol Langner at en.wikipedia.org - Originally from en.wikipedia; Licensed under Attribution via Wikimedia Commons - http://commons.wikimedia.org/wiki/File:David_Bohm.jpg#mediavi ewer/File:David_Bohm.jpg

p. 81: "John Stewart Bell (physicist)" from Queen's University Belfast – own work. Licensed under CC BY-SA 3.0 via Wikimedia Commons - http://commons.wikimedia.org/wiki/File:John_Stewart_Bell_(physic ist).jpg#mediaviewer/File:John_Stewart_Bell_(physicist).jpg

p. 84: "Hugh-Everett" from http://ucispace.lib.uci.edu/handle/10575/1060http://sites.uci.edu/ ucisca/2011/09/15/hugh-everett-iii-and-quantum-physics/ Licensed under Fair use via Wikipedia - http://en.wikipedia.org/wiki/File:Hugh-Everett.jpg#mediaviewer/File:Hugh-Everett.jpg

p. 86: "Double-slit experiment results Tanamura 2" from user:Belsazar - Provided with the kind permission of Dr. Tonomura. Licensed under CC BY-SA 3.0 via Wikimedia Commons - http://commons.wikimedia.org/wiki/File:Double-slit_experiment_results_Tanamura_2.jpg#mediaviewer/File:Double-slit_experiment_results_Tanamura_2.jpg

p. 87: "Buckminsterfullerene animated" from Sponk (talk) – own work, created with Pymol (0.99rc2) and Gimp (2.6.10). Licensed under CC BY-SA 3.0 via Wikimedia Commons - http://commons.wikimedia.org/wiki/File:Buckminsterfullerene_animated.gif#mediaviewer/File:Buckminsterfullerene_animated.gif

www.ingramcontent.com/pod-product-compliance
Lightning Source LLC
Chambersburg PA
CBHW020927180526
45163CB00007B/2913